Leonardo da Vinci

PRIM
PHYSICS

The principles behind Leonardo's science

Proudly supported by
The Science Foundation
for Physics

Marti Ellen
Illustrated by Andrew Davies

Introduction

Contents

First published January 2007
5th printing 2013
Published by
Sunshine Educational Pty Ltd
PO Box 551 Sans Souci
NSW 2219 Australia
www.primaryphysics.com
© 2011 Sunshine
Educational Pty Ltd

Primary physics,
the principles behind
Leonardo's science
Marti Ellen
ISBN 0-9586701-1-0

In order to appreciate the genius of Leonardo's science, it is important to study the "building blocks" which he used to come up with his extraordinary ideas. Leonardo made extensive studies of the components that make up machines. He did this in order to understand the principles by which they worked. His interest was beyond the particular machine which did this or that function. His focus was on understanding what he called "the elements of machines". Today we call these elements "simple machines".

Another physics principle which Leonardo incorporated was an understanding of energy and its effects. The storing and releasing of energy is a common theme in many of his machines, e.g. winches, springs, catapults, etc. He also incorporated a knowledge of forces: his plans for flying machines took account of gravitational forces; and in his file cutter machine, the heavy hammer is lifted by gears and is allowed to drop by gravitational force in order to make a cut. Frictional forces are lessened by the use of ball bearings, as in the machine for changing a theatrical stage, or are increased as in the drawing compass which has to have increased friction between the two legs in order to remain open with a constrant radius.

Leonardo also understood the properties of mass and the centre of mass (centre of gravity). He used weights and counter weights in numerous machines and he used an understanding of density and compressibility of air as the underlying principle of his aerial screw (a precursor to the helicopter).

Doing the lessons in this book will give you a hands-on understanding of these physics principles. Step by step, you will recognize what Leonardo did and why. At the end of the book, you can make working models of some of Leonardo's machines. And you will discover that the everyday events of our lives are operating under the laws of physics.

Acknowledgments We would like to acknowledge the help we have received, in so many ways, which has culminated in this book: Jan Hamilton, herself a physicist, who had the concept that physics could and should be taught to young children; Ken Dyers, whose communication training gave us the skills to be able to clearly communicate simple concepts; and to the many adults and children of the Kenja Social, Cultural and Sporting Association who have supported physics classes since 1985. Thanks to the many early students of Primary Physics who have gone on in science related careers. We thank Luigi Rizzo from Teknoart for the honour of including us in his inspirational Da Vinci exhibitions; and congratulate Tom Rizzo and his team on making the educational programme at the exhibition a big success. Thanks also to the many students currently using Primary Physics and the many more who will begin soon. Finally, a big thank you to Dick Collins for his scientific expertise, patience and support.   M.E., A.D.

In memory of our friend Ken Dyers 1922–2007

> **DISCLAIMER**
> The publishers and authors have made every reasonable effort to ensure that the experiments and activities in this book are safe when conducted as instructed but assume no responsibility for any damage caused or sustained while performing the experiments of activities in this book. Parents, guardians and/or teachers should supervise young readers who undertake the experiments and activities in this book.

The content of this book is an integral part of the educational program of the Da Vinci Machines Exhibition. Experiments were trialled and refined with the input of children over a number of exhibitions by the authors in consultation with Luigi Rizzo, the exhibition educational director. Reproductions of Leonardo's work in this book are from the exhibition *The Leonardo da Vinci Machines* organised by the Artisans of Florence – Teknoart. Used with permission.

The life of Leonardo da Vinci

Leonardo was born in 1452 near the town of Vinci during the Italian Renaissance.

He didn't have a formal education yet he taught himself throughout his life.

He learned to draw and write in his own way. His handwriting went backwards.

His drawings impressed the master Verrocchio enough to take him on as an apprentice painter and sculptor.

Verrocchio's studio completed the dome of Santa Maria del Fiore in Florence. Leonardo was fascinated with the clever use of machinery to raise heavy objects.

These experiences helped to broaden Leonardo's ideas of what was possible.

He studied things inside and out. He took careful measurements.

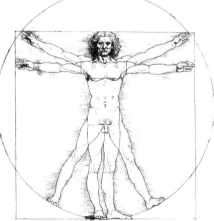

Through observation and experimentation he made many advances in every discipline in which he worked: he was an inventor, artist, anatomist, mathematician, engineer and musician. He was one of the first to use the modern scientific method.

Leonardo was a great artist. The *Mona Lisa* is the most famous painting in the world.

Leonardo's experiments were not always successful. The paints he experimented with for his masterpiece *The Last Supper* deteriorated soon after he completed it.

He was also a perfectionist and a procrastinator and left many projects unfinished.

Leonardo designed many things including flying machines, scuba diving equipment, armoured tanks and military weapons. Most were never built in his lifetime.

Leonardo's inventions were devised and drawn using simple machines that were understood in his time, such as levers, pulleys and gears.

He possessed a unique talent to invent things unknown in his day. Leonardo studied and mastered many subjects. His approach was the same in each: to understand the basic physical principles behind what happens.

section one
energy • mass • forces

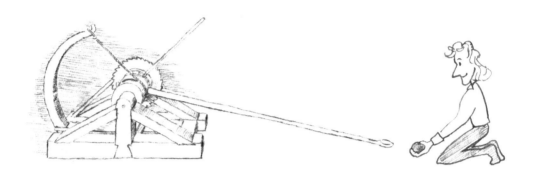

energy 1

Things you need

- plastic bag (e.g. shopping bag)
- lots of things to put in the bag (e.g. blocks, balls of different masses and sizes)

Words to use

faster
slower
more mass
less mass
energy

Extension

- Repeat the exercise using a large plastic garbage bag. For the mass, use some boxes of laundry powder. Compare your observations to the first exercise.

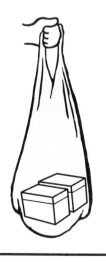

Storing energy

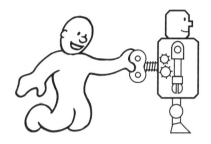

Releasing energy

Hold an empty bag by the handles with one hand.
Twist the empty bag 10 times.
Let go. What happens? _____

Put some mass in the bag

Twist the bag around

Let go

What happens now when you let go of the bag?

_____

Which has the most energy?

_____

energy 2

Thing

• mec
 con
 diffe

• san
 and

 OR

 OR
 (tha
 its s

Wor

pack
fill
full
pour
more

Exte

• Bu
 mc
 stru

• Bu
 ob
 (e.

Things you need

• folding table OR large board
• toilet rolls (empty)
• marbles
• plastic bag

Words to use

fast	slow
up	down
slide	roll
ramp	distance
speed	measure

Extension

Compare the energy of different ramps (e.g. steeper, shorter).

Storing energy

Releasing energy

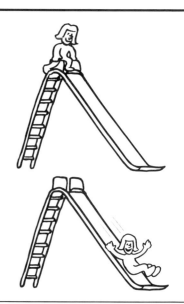

What happens when you let go of the toilet roll? (Draw an arrow)

Wrap marbles in a plastic bag and pack them into another toilet roll. Let go of both rolls at the same time.

Which has the most energy?

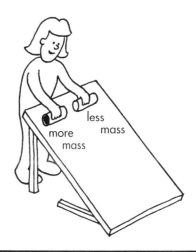

less mass

more mass

centre of mass 1

PRIMARY PHYSICS *The principles behind Leonardo's science*   20

Things you need

- dolly peg
- wire, 1 mm (1/16") dia.
- wire cutters
- 2 washers
- plate
- hardcover book

Words to use

balance   middle
centre   mass

Extension

- You can also have fun decorating the toys.

Keep this for use in the next experiment.

Balance

Make this balancing toy.
Adjust the wires so that the peg will stand vertically.

The **centre of mass** is the point at which it balances.

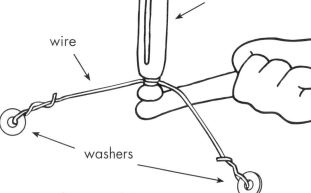

wire

dolly peg

washers

Colour in the centre of mass of this toy.

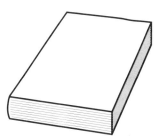

Balance these objects on your finger.
Mark an **X** on the centre of mass of these shapes.

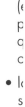

centre of mass 2

Things you need

- balancing toy from last lesson

Words to use

centre of mass
balance
underneath
middle

For each of the shapes below, locate the centre of mass by finding where it balances on your finger.

1.

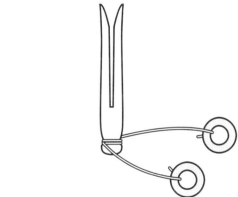

2.

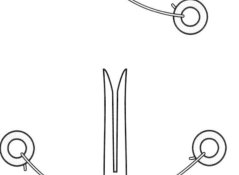

Extension

- Keep changing the shape, and find the new location of the centre of mass.
- Locate the centre of mass of other objects.

3.

Draw an **X** where you put your finger.

centre of mass 3

Things you need

- balance scales
- 2 buckets
- pairs of objects of equal mass (e.g. blocks, shoes, marbles)

Words to use

balance   mass
bigger   smaller
scale

Extension

Use masses of known weight, (e.g. tinned food).

balanced	unbalanced

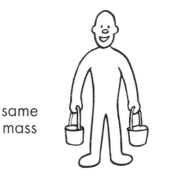

same mass — same mass

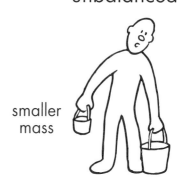

smaller mass — **bigger mass**

Use your own objects to make a **balance** look like this:

1. balanced

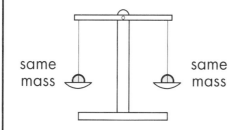

same mass — same mass

2. unbalanced

smaller mass — **bigger mass**

Draw a **balanced** scale. Draw the objects you used in the buckets.

Draw an **unbalanced** scale. Draw the objects you used in the buckets.

centre of mass 4

Things you need

- balance OR
  number line OR
  ruler

- objects to use
  as weights,
  e.g. rubbers,
  coins, plasticine
  (modelling clay)

Words to use

balance	centre
more	less
mass	equal

Extension

By experimenting
find out what the
rule is that keeps
the centre of mass
at zero.

unbalanced   balanced

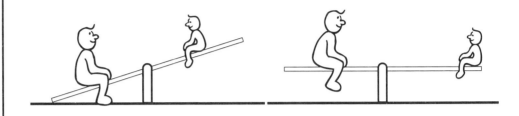

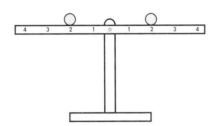

Put equal masses the same
distance from the centre of
each side.
Mark an **X** to show where the
centre of mass is.

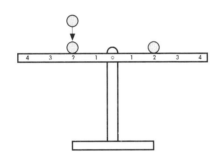

Add mass to one side.
What do you see?

Add mass to the other side
to make it balance again.

Try moving the smaller mass
further out to make it balance
(without adding more mass).

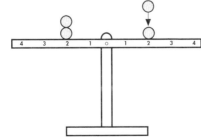

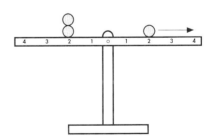

gravity and mass 1

Things you need

- wooden board
- flat head nail
- very flexible rubber band OR elastic string
- 3 equal rubber bands
- paper clip
- plastic container OR plastic cup (punch 3 holes with a hole punch in advance if possible)
- objects of known weight (e.g. 50 g, 100 g, 200 g (2 oz, 4 oz, 8 oz))
- scissors OR hole puncher to **carefully** make holes in the cup
- hammer

Words to use

mass	gravity
pull	exactly
more	less
between	close to

Extension

Use different combinations of masses to further calibrate the gravity machine.
(See next page.)

Give the cup a pull so that the rubber bands are taut. Then let go before marking the 'zero' line.

Make something to measure the pull of gravity.

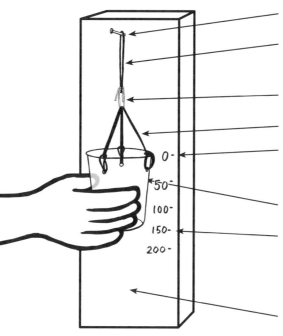

- nail
- long, thin, flexible rubber band OR elastic string
- paper clip
- 3 rubber bands of equal length
- Mark the zero line on the wooden board level with the top of the cup.
- plastic cup (with three holes)
- Calibrate by placing known masses in the cup and mark 50 g, 100 g and 200 g on the wooden board
- wooden board

0 -
50 -
100 -
150 -
200 -

What force makes your machine work? _____

gravity and mass 2

Things you need

- gravity machine from previous lesson
- golf ball
- marbles
- potato
- hard-boiled egg
- onion
- apple
- orange

Words to use

mass   bigger
smaller   scale

Try your machine with different amounts of mass.

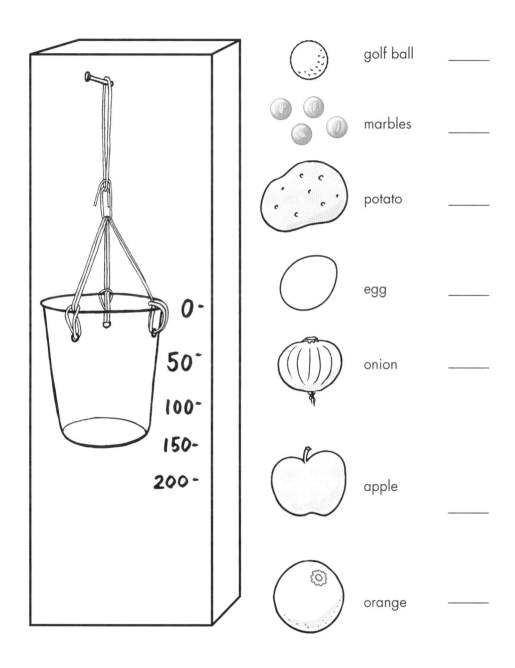

golf ball   _____

marbles   _____

potato   _____

egg   _____

onion   _____

apple   _____

orange   _____

Extension

How many marbles does it take to make 100 g (4 oz)?

Draw a line from each object to the point it measures on the scale. Write the approximate mass on the line next to the picture.

gravity and mass 3

Scales measure weight

A scale measures the amount of gravity pulling on the mass.

Use a scale to measure how heavy things are.

Draw a scale.

Fill in the chart

object	weight
book	

gravity and mass 4

Things you need

- 2 identical small containers e.g. empty 250 ml (8 oz) fruit cups
- sugar
- white flour
- a ruler and a pencil to make a scale

Words to use

more mass
less mass
level   balance
equal   same
weight   density

Same size, different mass

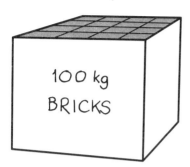

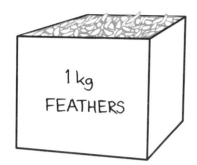

Fill the two containers. Level the top with a ruler.
Put them on the scale to see which has more mass.

flour   sugar

Which one has the most mass? _____

Can two things have the
same size but a different mass? _____

Extension

Which way must you move the pencil so that the two masses balance?

At what number on the ruler does it balance?

density 1

Things you need

- paper
- aluminium foil

both cut to same size

Words to use

more   less
packed space
size   large

Extension

- Try wetting the paper to see if it can be packed tighter.
- Try comparing a very thin sponge to wet paper, by crunching them both as small as possible.

Packing in

How small can you make a piece of paper?

How small can you make the same sized piece of aluminium foil?

Draw them.

Which one can be packed the most tightly or **densely**? _____

density 2

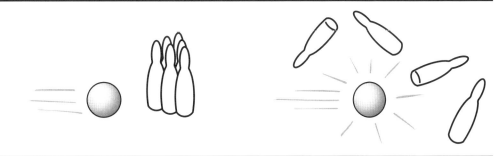

Things you need

- lots of marbles
- strip of plastic, string or tape
- large marbles OR even bigger balls (e.g. golf ball or cricket ball)

Words to use

close   packed
tight   energy
hit   expand
move   space
more   less
circle   centre
scattered
spread apart

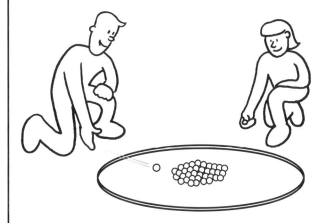

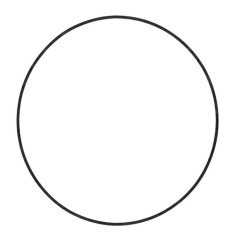

Everyone sit around a circle, taking turns shooting large marbles into the central pile of marbles.
Let the marbles stay where they land.
Do this 20 to 30 times.

Extension

Piñata!

Have a game of piñata. Compare the density of the contents before and after it breaks.

Draw the pile of marbles at the start.

Draw the pile of marbles after shooting at them.

Which pile of marbles is more dense?

density 3

Things you need

Pieces of:

- marshmallow
- banana
  (slice to size of
  marshmallow)
- white bread
- pumpernickel
  bread

Words to use

more   less
firm   soft
dense

Chew each of the foods.

banana

pumpernickel bread

marshmallow

white bread

Which is the most dense? (Circle one)

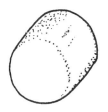
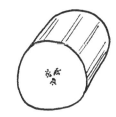

1.   marshmallow   OR   banana

2.   pumpernickel bread   OR   white bread

Extension

- Try other
  combinations:
  fresh versus dried
  fruit (e.g. sultana
  & grape; dried &
  fresh apricot).

review

 You can store up energy and release **energy**.

Every thing has **mass**. Mass takes up **space**.

 When more mass is in a space it has more **density**.

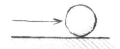 A **force** moves mass.
The bigger the mass, the more
force it needs to move.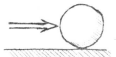

Friction slows it down.

 Gravity makes things fall.
The force of gravity on mass gives it weight.

 The **centre of mass** is where it balances.

quiz

1. **Which sentence goes with which picture?**

   Point to the right answer with an arrow.

   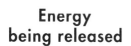 **Energy being released**

   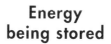 **Energy being stored**

   No energy being used

2. **Which container holds the most?** Circle the right answer.

   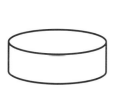 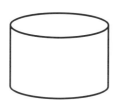

3. **Which one is easiest to pull?** Circle the right answer.

    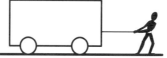 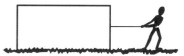

3a. **What is the name of the force which makes it harder to pull along?** Circle the right answer.

   gravity   **friction**   **density**   **mass**

4. **Which will hit the ground first?** Circle the right answer.

    the brick

    the tissue box

    both objects together

    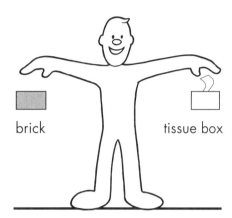

    brick   tissue box

5. Where is the centre of mass on each of these objects?

    Mark centre with an **X**.

6. Which one weighs more?

    Circle the right answer.

6a. How much more?

    _____

    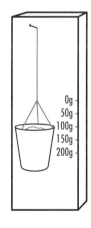

    0g
    50g
    100g
    150g
    200g

    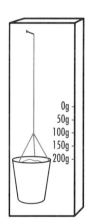

    0g
    50g
    100g
    150g
    200g

7. Which is most dense?

    Circle the right answer.

puzzle

balance   friction
centre of mass   gravity
density   mass
energy

Fill in the missing letters.

e _ _ _ _ _ _

g _ _ _ _ _ _ _

f _ _ _ _ _

f _ _ _ _ _ _ _ _

m _ _ _ _

d _ _ _ _ _ _ _

b _ _ _ _ _ _ _

c _ _ _ _ _ _ of m _ _ _ _

section two
the simple machines

Simple machines make moving loads easier.

easier to move along

easier to transfer movement

easier to go downhill

easier to go around than straight up

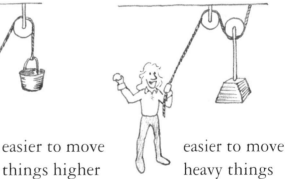

easier to move things higher

easier to move heavy things

easier to pick things up

easier to turn things

easier to balance things

wheels 1

Things you need

- 1 block of wood approx. 8cm x 10 cm x 20 cm (2" x 4" x 8")
- 4 equal sized metal lids
- 8 washers
- 4 short nails
- 4 rubber bands
- ruler
- hammer (be careful not to hit your fingers when hammering)

Words to use

roll	faster
rubbing	slower
friction	energy
gripping	axle
turn	

Extension

Build a cart using other materials for wheels. e.g. old training wheels, wheels from Mechano™ sets or other building sets, wooden discs, discs made of other materials including rubber.

How well does this cart move compared to the cart with metal wheels?

Try letting the cart go down a ramp. How far can it go?

Build a cart

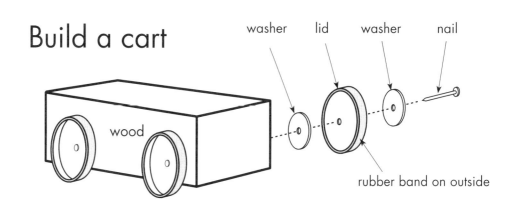

washer   lid   washer   nail

wood

rubber band on outside

Work out the centre by finding the diameter. Punch a hole right in the middle of the lid.

Don't pound the head of the nail into the lid. The lid must be able to rotate freely.

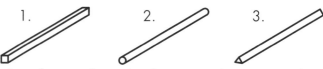

Make sure the nails (axles) are all even and that the wheels all sit flat.

The nail acts as an **axle** to let the wheel turn.

The wooden block is the **load** which is moved by the wheels.

Push the cart along the floor or roll it down a ramp.
If the wheels don't turn easily, make adjustments so they do.

Which would be good to use as an axle:
(Circle your choice)

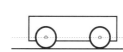

1.   2.   3.

Give three examples of ways that wheels are used to move a load.

1. _____

2. _____

3. _____

Wheels help us move _____ .

wheels 2

Things you need

- blocks of wood
- long nails
- hammer
- empty cotton spools
  (OR plastic spools available at toy shops or craft shops)
- large rubber bands

Words to use

friction   rubbing
slower   faster
pull

Extension

Try other combinations, for example:

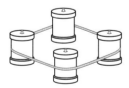

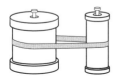

Build a drive belt.

Attach two spools with one rubber band. Turn one.

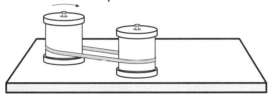

Note: don't pound the nail in while the spool is on the nail as it may get damaged by the hammer.)

What happens? _____

Attach more than one together. Turn the first one.

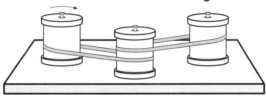

What happens? _____

Twist one rubber band. Turn one spool.

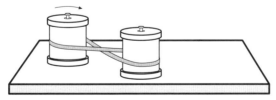

What happens? _____

What is the load in these drive belts?:

1) supermarket check-out _____. 2) escalator _____.

wheels 3

Things you need

- block of wood
- short nails
- bottling tops ("crown" bottle tops)
- hammer

Words to use

turn   cog
wheel   axle
gear

Extension

Turn the pedals on a bike.

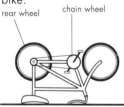

rear wheel   chain wheel

Which way does the chain wheel turn?

Which way does the rear wheel turn?

How are the chain wheel and rear wheels connected?

Which wheel turns more quickly?

How many times as quickly?

If your bike has different speeds, keep the chain on the same gear at the back and switch the front gear from the larger to the smaller several times.

On which gear does the back wheel turn fastest: smaller or larger?

gears

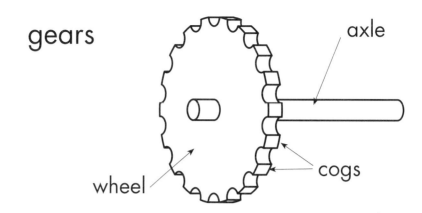

axle

wheel   cogs

Build some gears

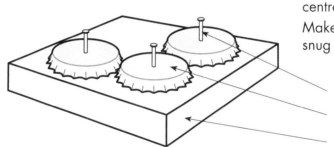

Important tips

The nail must be right in the centre of the top.

Make sure the bottle tops are snug against each other.

nail

bottle tops

wood

If you turn gear 1, which way does gear 2 turn? *(Draw an arrow)*

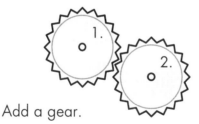

Add a gear.

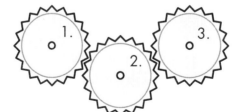

Which way does gear 3 turn? *(Draw an arrow)*

An example of a machine which uses gears is a clock.
What is the load which is moved by the gears? _____

pulleys 1

Things you need

- bucket
- pulley
- short rope
- long rope
- water (or weights)
- climbing bars (or monkey bars)

Words to use

raise   lower
force   friction
gravity   higher

Be careful to stand behind the line.

Extension

Have relay races lifting the bucket up to the top a certain number of times without spilling the water.

Excursion

Indoor rock climbing:

Measure how high you climb by measuring the length of rope pulled.

Note how many pulleys are used.

A **pulley** is another use for a wheel and axle.

Load

Set up a pulley. Use a climbing bar, rope, pulley and bucket filled halfway with water.

Draw a safety line on the ground. Everyone has to stay behind the safety line until their turn.

Note: **Be careful! No one should be underneath the bucket.**

Try lifting the bucket up to the bar using just your arm.
Can you do it? _____

Now use the pulley to raise the bucket to the bar.

Can you do it? _____

Is it hard work? _____

Give everyone a chance to lift the bucket up as high as possible 5 times in a row using the pulley.

When you pull downwards on the pulley, which way does the bucket move? _____

Safety note: When lowering the bucket, *slowly* move your hands "hand over hand".

pulleys 2

Things you need

- 3 pulleys
- 2 ropes, 3 m (3 yards) each
- 2 buckets
- water (OR weights)
- short pieces of rope (to tie the pulleys to the railing and bucket)

Words to use

pull	longer
force	shorter
fixed	further
work	moveable
easier	harder

Extension

Where do you see pulleys used in your home?

What are they used for?

Draw a picture of your pulleys.

Try this

Tie a rope to one broom handle and wrap the rope around both several times. Hold on to the free end. Ask your friends to pull the sticks apart, while you pull on the rope. They won't be able to. You've made a very efficient pulley system.

Set up a single and a double pulley.

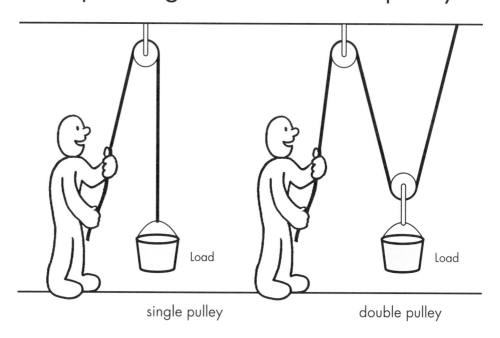

single pulley   double pulley

Put the same amount of water (or weight) into each bucket.

Have a turn lifting each bucket.

Which is easier? _____

What are some advantages of using a double pulley? _____

_____

What are some disadvantages of using a double pulley? _____

_____

Colour the fixed pulleys *red*. Colour the moveable pulleys *blue*. Colour the load *green*.

pulleys 3

Things you need

- single and double pulleys (from the previous lesson. Set up in advance if possible.)
- pencil (to copy down data)
- tape measure
- marking pen

Note

To measure the length the rope was pulled:

1. Pull the rope so that the bottom of the bucket is at the designated height. Be sure the rope is taut and vertical. Mark the rope clearly where it touches the floor.

Take care to stand aside

2. Lower the bucket to the ground and again mark the vertical rope where it touches the floor.

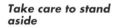

3. Measure the distance between the dots.

Measure how far you need to pull the rope to lift the bucket.

Single pulley

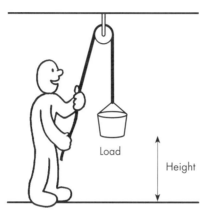

Load

Height

Height bucket pulled	Length rope pulled
30 cm (1')	
60 cm (2')	
90 cm (3')	

Double pulley

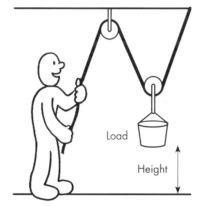

Load

Height

Height bucket pulled	Length rope pulled
30 cm (1')	
60 cm (2')	
90 cm (3')	

What pattern did you see for the single pulley? _____

_____

What pattern did you see for the double pulley? _____

_____

What conclusion can you draw from these results? _____

_____

ramps

Things you need

- toy car
  (a Matchbox™
  car will do
  OR use your cart
  from **wheels** 1)
  OR marbles
- plank (OR a table
  with folding legs)
- block (OR chair)
- measuring tape
- ruler

Words to use

roll	farther
ramp	shorter
faster	slower
gravity	easier

Let your car roll down the ramp.

Measure the distance it rolls. Fill in the table (**first height** & **distance 1**).

Using the same car, repeat twice, filling in the table (**distance 2** & **distance 3**).

height

distance

Try this at a greater height and a lower height. Make sure the highest one is not so high that the car stops when it hits the floor.

height	distance 1	distance 2	distance 3
first height			
greater height			
lower height			

What is the job of the ramp? _____

The greater the height, the _____ the car rolls.

The lower the height, the _____ the car rolls.

Why is this? _____

wedges

Things you need

- large zipper (open-ended)
- paper & pencils
- glue stick
- nails

Words to use

focus   point
wedge   along
hammer   centre
meet   forces
concentrate

Extension

Only do this with adult supervision OR get an adult to do this as a demonstration

• Use a hammer and chisel to make wood shavings from a large log. Go with the grain.

See if you can fill up a bag full of wood chips.

Do you think the chisel and hammer make it easier to split the wood?

_____

Does the shape of the chisel help you?

_____

How?

_____

_____

• Have an adult demonstrate splitting logs with an axe.

A wedge is a double-sided ramp that can be used to force something apart.

axe

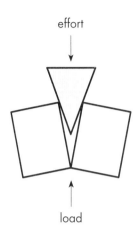

effort

load

Trace this ramp twice and cut the pieces out

Glue them here to make a wedge.

Examine a zipper.

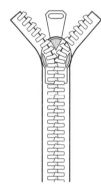

How does it work?

_____

_____

_____

A nail is another type of wedge.

Look at a nail. How many sides make up the wedge at the sharp end?

_____

What other wedges can you name? How many sides are they made of?

_____

_____

_____

screws 1

Things you need

- paper,
  10 cm x 10 cm
  (4" x 4")
- a pencil
- coloured pencils
- scissors
- sticky tape

Words to use

screw	up
spiral	down
wind	turn

Extension

Using some small pieces of soft wood, some screws, and screwdrivers, practice screwing into wood.

You may need to start with small pre-drilled holes, or use a hammer to start the screw.

What is the force which holds the screw in the wood?

_____

water slide

screw

Fold a square piece of paper to make a triangle.

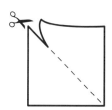

Cut out the triangle.

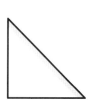

Colour in a stripe along the edge.

Wrap the paper around a pencil.

Tape the end.

Start at one end and follow the coloured stripe with your finger around the new shape. Where does it go? _____

What is the shape of the coloured stripe? _____

screws 2

Things you need

- large nut and bolt
- cardboard
- scissors OR something to punch a hole in the cardboard

Words to use

up   extend
down   work
spiral   lift
lower

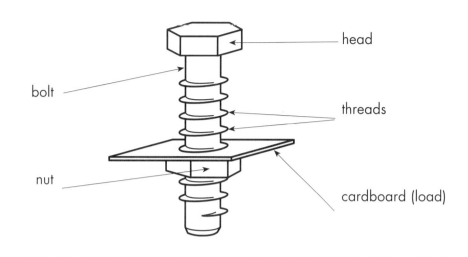

head

bolt

threads

nut

cardboard (load)

Screw the nut up and down.
What happens to the cardboard? _____

_____

Describe the path of the
threads on the bolt? _____

_____

Name an object or machine
which moves something up
and down in this way. _____

Extension

Examine a swivel chair or stool (a screw one not a pneumatic-powered one).

Observe the way it moves.

Examine a plastic
container and its lid.
How does it work?

Fit the lid on the
container.

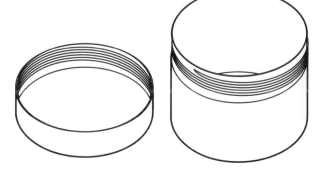

levers 1

Things you need

- 30 cm (12") ruler
- pencil
- counter (or coin)

Words to use

fulcrum   force
load   effort
middle   far
close

_____

_____

_____

_____

_____

See-saw

Balance a ruler on a pencil to make a lever.
Place the pencil at 15 cm (6"). Place a counter on the 0 mark.
Press on the other side of the ruler to make the counter flip up.

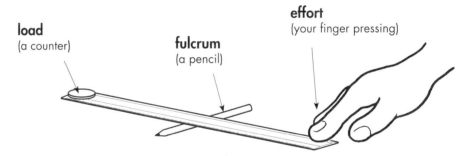

load
(a counter)

fulcrum
(a pencil)

effort
(your finger pressing)

Next place the pencil at 7 cm (3"); then at 23 cm (9");
and try again.

Extension

At the playground, use a see-saw to test out the best place to put a heavy load in order to lift it.

bucket full of sand

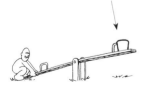

Be careful not to hop off without telling your friend.

When we want to make the movement of the load big, the effort must be closer to the fulcrum than the load.

When we want to cause a big force, the effort must be farther from the fulcrum than the load.

Where do you place the pencil to make the counter go:

- up the highest? _____
- up second highest? _____
- up the lowest? _____

Which gives the biggest push?:
(a) when the effort is close to the fulcrum; or
(b) when the effort is far from the fulcrum   *Circle* (a) *or* (b)

levers 2

Things you need

- wheel-barrow
- a flat space to use the wheel-barrow

Words to use

fulcrum	work
load	tip
effort	stable

Wheel-barrow

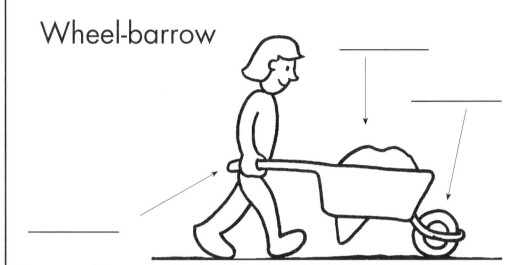

Using a wheel-barrow, lift some heavy things and carry them for 25 m (25 yards).

Adult supervision.
You can carry each other *if you're careful.*

Try carrying them the same distance without the wheel-barrow. Which is easiest? _____

Where are the load, fulcrum and effort on the wheel-barrow. (Label the top drawing)

Where is the safest place to put the load in the wheel-barrow when it is standing by itself (and no-one is holding on to the handles)? _____

Why? _____

Extension

What would happen if the load was near the handles.

What happens to the fulcrum?

_____

levers 3

Things you need

- chopsticks
- rubber band
- scrap paper
- noodles
  (e.g. macaroni,
  shell, etc.)

Words to use

fulcrum
load
effort

Extension

- Have two people
hit a ball to each
other with tennis
racquets (or other
bats, paddles, etc.)

Observe them.

Describe how the
body and racquet
together work as a
lever.

- Dig a hole with a
shovel. Where are
the fulcrum, load
and effort on the
shovel.

Chopsticks

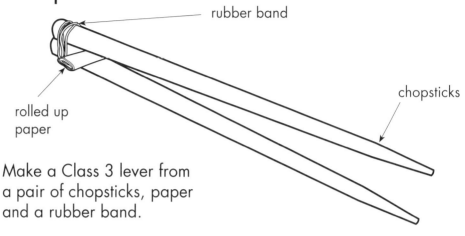

rubber band

rolled up
paper

chopsticks

Make a Class 3 lever from
a pair of chopsticks, paper
and a rubber band.

Eat a bowl of noodles with the chopsticks.

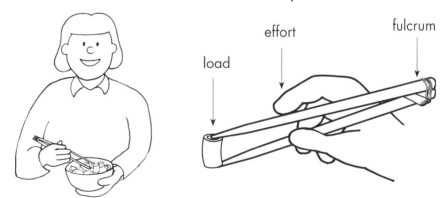

load

effort

fulcrum

What other lever are you using to chew the noodles? _____

_____

Name another Class 3 lever in your body and how you use it.

_____

_____

levers 4

Things you need

- tongs
- see saw
- garlic press
- nut-cracker
- scissors
- stapler
- scales

Words to use

lever   class
fulcrum   type
load
effort

Extension

Make a list of other levers you have seen, what they are used for, and what is their class of lever.

_____

_____

_____

_____

_____

_____

There are 3 basic kinds of levers

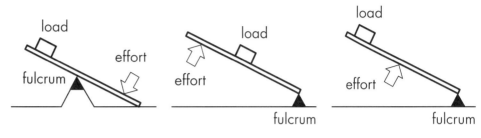

class 1 lever   **class 2 lever**   **class 3 lever**

Which class of lever are each of these things?

tongs

Class ____

garlic press

Class ____

scales

Class ____

scissors

Class ____

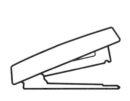

stapler

Class ____

nut-cracker

Class ____

levers 5

Things you need

- scissors
- scrap paper
- nut cracker
- nuts in shell (e.g. walnuts)
- tongs
- marbles
- 2 plastic containers

Words to use

fulcrum
load
effort

Extension

Have a party.

Use different types of levers for preparing and serving the food. List the utensils you use below.

_____

_____

_____

_____

_____

"Give me a firm spot to stand on and I will move the earth."

– Archimedes
c.220 B.C.

Set up three stations with the three types of levers. Try each.

Class 1
Cut out a shape from a piece of paper using scissors.

Class 2
Crack open some nuts.

Class 3
Use tongs to transfer marbles from one container to another.

levers 6

Things you need

- cardboard,
  60 cm x 10 cm
  (24" x 4")

- stiff cardboard,
  20 cm x 20 cm
  (8" x 8")

- 2 paper fasteners

- string,
  80 cm (30")

- single hole-
  puncher

- scissors

- stapler, staples

- ruler

Make moveable ears.

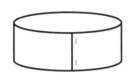

1 Measure a strip to fit around your head.

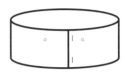

2 Staple the ends together.

3 Punch two holes about 10 cm (4") apart, 2 cm (1") from the edge of the cardboard.

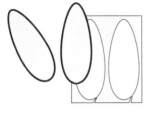

4 Cut out two ears from the stiff cardboard.

5 Punch two holes at the end of the ears (don't punch holes too close to the edge).

6 Attach the ears with paper fasteners through the bottom holes.

7 Attach ears to each other with a short piece of string through the top holes.

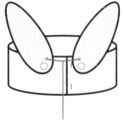

8 Tie a long string in the centre of the short string.

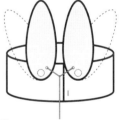

9 Pull on the long string to make ears move up and down.

Extension

Make up a play or a song in which everyone moves their ears at the same time.

Use an arrow to point the word to the part on the picture.

What type of lever is this? _____

load

fulcrum

effort

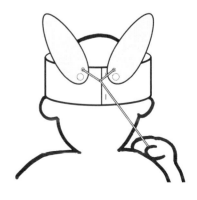

tools

Things you need

- egg beater
- ice cream scoop
- pizza cutter
- corkscrew

Be careful of any sharp objects

Words to use

screw   ramp
wheel   wedge
pulley   lever
gear   drive belt
load   lift
roll

Extension

Find other objects that incorporate simple machines.

List them.

Everyday machines

Examine objects to find out how simple machines are used to do everyday jobs.

object	type of simple machines making up each tool	number of simple machines	how they contribute to moving a load
egg beater			
ice cream scoop			
pizza cutter			
corkscrew			

section three
working with Leonardo

Before Leonardo could be innovative, he had to understand;
and to do that, he studied what already existed.

A wheel
helps things to move

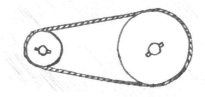

A drive belt
can help another wheel to move

A pulley
makes lifting
easier

A wedge drives
things apart

A screw
makes a
longer but
easier path

A ramp makes it easier to *move up or down*

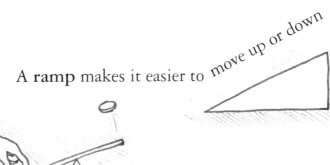

A lever reduces the amount of effort to move something

project 1 robot arm

Things you need

- 3 cardboard strips,
  30 cm x 5 cm (12" x 2")
  (use stiff cardboard)
- 2 paper fasteners (brads)
- one-hole punch
- wire coat hanger
- paper clip
- set of old keys on a key ring
- scissors
- ruler

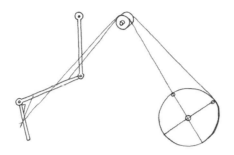

Here is a sketch of Leonardo's robot arm. Astronauts work on space shuttles with a robot arm similar to this design.

1 Punch a hole in each end of the three cardboard strips.

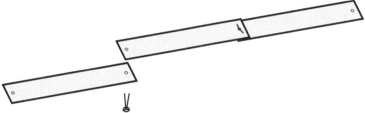

2 Use paper fasteneners to connect them. Be sure they can move freely.

3 Untwist the top of a coathanger and straighten out the hanger. Make a small hook at the end.

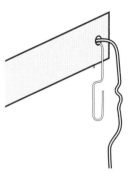

4 Unbend a paper clip to make a hook.

5 Insert the paper clip hook in the same hole as the coat hanger.

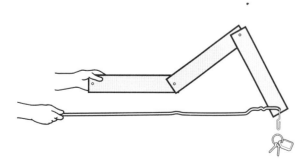

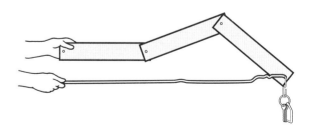

6 With one hand stationary and the other hand manoeuvering the wire, move the arm so you can pick up the keys.

7 Replace the keys on to the table.

project 2 ball bearings

Things you need

- 2 metal lids, 7 cm (2¾") dia.
- 10 marbles (of equal size)
- 5 washers, 2.5 cm (1") dia.
- 1 bolt & nut, 5 cm (1")
  x 2 mm (¹/₁₆") dia.
- 1 nail, 2 mm (¹/₁₆") dia.
  (to make holes in lids)
- hammer
- electrical tape
- scissors

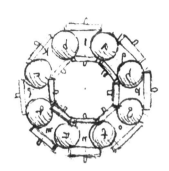

Here is a sketch of Leonardo's drawing for building a machine to move theatre stage sets. He uses ball bearings in order to minimize friction.

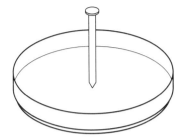

1 Find the centre of the lids and make a hole with a nail. (If the hole is too small, wiggle the nail around to make the hole slightly larger. Cover the hole with tape to avoid burrs, then make a new hole through the tape.)

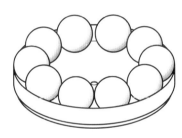

2 Place 10 marbles around the perimeter of the lid.

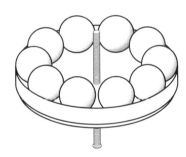

3 Insert the bolt through the centre hole.

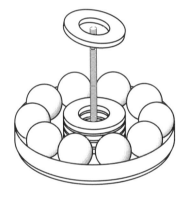

4 Place a stack of washers over the bolt and inside the ring of marbles.

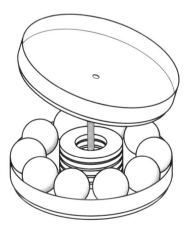

5 Place the second lid on the bolt, facing up (like the first lid).

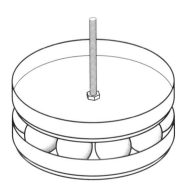

6 Use the nut to tighten the two lids together.

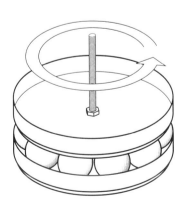

7 Check that the lids swivel easily over each other.

project 3 revolving crane

Things you need

- ball bearing model from project 1
- 1 bolt, 5 mm ($\frac{1}{8}$") dia.
- 6 paddlepop (popsicle) sticks
- 4 headless matchsticks
- plastic lid, at least 12 cm (5") dia. Tough and strong.
  Adults only: Make a small hole in the centre of the lid beforehand.
- 2 clothes pegs
- craft glue
- rubber band

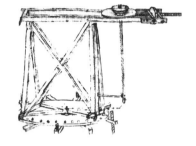

Here is a sketch of Leonardo's drawing for a revolving crane. A similar crane was used to add the ball to the top of the dome in Florence.

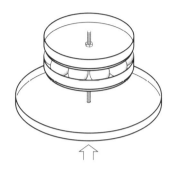

1 Carefully remove bolt from your ball bearing model; then re-insert including a lid.

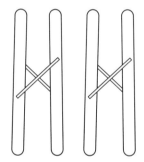

2 Glue 2 match sticks to each side of 2 paddlepop sticks. Make 2 sets.

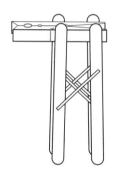

3 Glue a set to each side of a clothes peg.

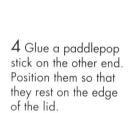

4 Glue a paddlepop stick on the other end. Position them so that they rest on the edge of the lid.

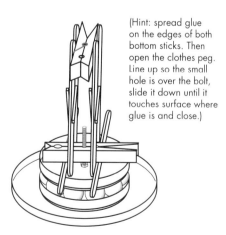

(Hint: spread glue on the edges of both bottom sticks. Then open the clothes peg. Line up so the small hole is over the bolt, slide it down until it touches surface where glue is and close.)

5 Glue a clothes peg across the two bottom sticks. The small hole must line up over the bolt.

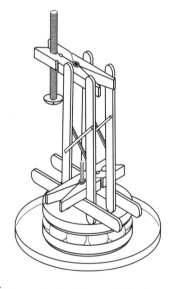

6 Screw the bolt through the large hole in the top clothes peg.

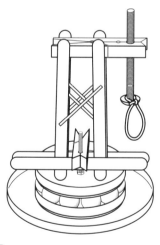

7 Tie a knot in a rubber band and join it to the bolt as a carrier. You can load something into the carrier and move it.

project 4 invent your own

Things you need

- anything you want to use. e.g.:

  wooden boards
  wood
  paddlepop
  (popsicle) sticks
  nails
  screws
  bolts
  hinges
  bottle tops
  lids of jars
  toilet rolls
  string

Things and words to use

  wheel
  gears
  drive belt
  pulley
  ramp
  wedge
  screw
  lever

Design and make something of your own using at least three types of simple machines.

Draw your model here. Write the name of each simple machine and show where it is on your drawing.

outcomes and helpful hints

Please read the section on each experiment before doing it, as there are many helpful hints.

 energy

An introduction to the scientific methods of observation, experimentation, and the carrying out of set procedures.

energy 1 Do this simultaneously – it is more fun and more dramatic! Everyone should hold the handles of their bag in one hand and twist ten full turns with the other hand. The potential energy of the twisted bag increases with greater mass put into the bag. This is observed as the greater kinetic energy of the spinning bag. The relationship of *more mass → more energy* is the point we wish to make.

energy 2 The toilet roll filled with marbles rolls faster. The *more mass → more energy* relationship holds. Note, however, that this is mainly due to the air resistance against the empty cardboard roll, and not that a heavy thing falls faster than a light one (see gravity 2.) Emphasise the *more mass → more energy* concept, rather than *air resistance* which is explained in gravity 3.

 mass, space, volume

This section looks at mass ('physical things') and the space it occupies. It also introduces simple measurement. The emphasis is on exploring.

 forces, friction, gravity

These experiments introduce changing and moving mass. The **forces** experiment demonstrates three things:

1 left on its own the mass will not move; (apart from some very subtle flattening of the dough).

2 to push the mass along the table, one feels the friction between the table and the dough slowing the movement; and

3 when it goes over the edge, gravity takes it to the floor.

Friction (the rubbing force) is best first demonstrated by getting the students to rub their hands together and feel the heat generated by the friction.

Friction 1 shows the effect of friction on an object. **Friction 2** The extension section is very informative. It shows that **contact** is important. Friction is a force that depends on contact.

Gravity (the falling force) is actually the attraction between two masses: the object and the planet, where, due to the vast differences in the amount of mass each has, the object appears to 'fall' to the earth.

In **gravity 1** we observe that the longer an object falls the more its velocity increases.

This is evidenced by the increasingly dramatic effect on the object when it collides with the planet. It has more kinetic energy, therefore more energy is dissipated as the motion stops. This is demonstrated by the distortion of the shape of the dough.

gravity 2 demonstrates that objects fall at the same rate, regardless of the difference in their mass. The two objects have to be released at the same time from the same height to be seen (and heard) hitting the ground simultaneously. But don't worry if it isn't exact, even Galileo had to deal with this. Contrast the two marbles with a marble and a tissue (gravity 3). (By the way, the longer an object falls, the faster the gravity of the earth makes it fall – at a rate of $9.8 m/s^2$.)

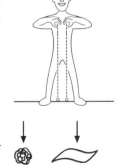

gravity 3 shows something to watch out for. It is not that the law of gravity is suspect, simply that the shape and mass of some objects lead them to be slowed down by the friction of the air. In the first extension exercise, the tissue falls at the same rate as the book. The book pushes the air out of the way.

outcomes and helpful hints

 ## centre of mass

This section looks at the centre of mass (sometimes referred to as the centre of gravity), which   is the spot in the middle of any given object where it balances. The point is not always necessarily inside the object itself as many of these experiments show.

In making the balancing toy used in the first two experiments, the peg and the two  arms should all lie in the same plane. Then it must be carefully balanced on the finger with some patience. It may take some adjustment to the toy before the peg stands up vertically.

By changing the shape of an object we change the position of its centre of mass. By moving the wires we change the centre of mass and the peg tilts. Soon the children will be able to predict the location of the centre of mass.

In centre of mass 3, we compare the pull of gravity on objects of different size. Accuracy becomes even more important in the next experiment where we start to work out what makes balance. A scale balances when the centre of mass is at the pivot point. Two ways to achieve this are explored in centre of mass 4: one can either have equal amounts of mass the same distance from the pivot creating a symmetrically shaped balance scale (3), or change the shape by redistributing the mass (4). e.g. two balls at '2' balances with one ball at '4'. This is the principle of a 'lever' and a prelude to levers 1 of simple machines in section two.

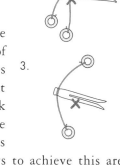

(3) 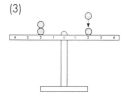 (4)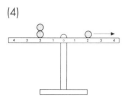

 ## gravity and mass

Mass is affected by gravity. We measure the pull of gravity on mass and call it weight. Mass and weight are not the same thing, though the words are often mistakenly used interchangeably due to the consistent effect of gravity on earth. At this stage we won't talk of 'heavier' or 'lighter' rather that it has 'more mass' or 'less mass'.

We begin by making a simple scale to show that the more mass an object has, the greater the effect of gravity. Following that we move on to conventional weighing devices which operate by principles which are then understood by the student. At this stage we can introduce the concept of weight, the measure of the pull of 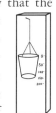 gravity on an object. The greater the measured weight, the greater the mass must be. If we were to use the same equipment on the moon where the gravity is less (due to the moon having less mass than the earth) we would find the same objects weighed less even though their mass remained the same.

In the final experiment in this section we compare two objects of the same size but different mass. The objects' masses are different due to their differing densities. We do this experiment as a prelude to looking at the concept of 'density'. You may need to refer back to this experiment in the next section to be sure the concept is clear.

density

In the first experiment we get the idea of different densities, and the idea of density being mass packed into a space.

density 2 refines this concept with a picture of 'particles' distributed in space. The extension experiment is a fun demonstration, but it illustrates the point. The densly packed contents are dispersed with a sudden great energetic whack!

density 3 provides an enjoyable first-hand experience of density.

outcomes and helpful hints

 wheels

wheels 1 **Build a cart.** Younger children will need to practice using a hammer. Get some extra wood (free offcuts from hardware stores) and nails. It will prove very productive to practice hammering skills, as this is part of several of the exercises. Small hammers are a wise idea. *Helpful hint:* Use the nail to make the hole in the lid. Then place heavy tape over it to cover the burr. Poke the hole again through the tape.

Check whether the hole is too small for the wheel to rotate freely by inserting the nail. If it is too tight, rotate the lid on an angle so the hole gets bigger. Be sure not to pound the nail in too far.

• The rod that is a circle in cross-section is the best choice for an axle.

• Wheels help us move loads.

wheels 2 Uniform-sized colourful plastic spools can be purchased at craft shops or toy stores.

A great place to see an application of the drive belt is in an old tape recorder or the checkout counter of a supermarket.

wheels 3 When nailing through the bottle-tops, be sure to hit the nails and not the bottle-tops. If the tops are hit too often they lose their shape and this will affect the way they turn. Keep tops as close together as possible, whilst remaining flat against the wood. This is when hammering practice becomes important.

If you have some very keen children, have them make a mark on the rear wheel and chain wheel of the bicycle. Ask them to work out how many times the chain wheel goes around for each turn of the big rear wheel.

pulleys 1 Set this one up at the swings in a playground. In this exercise you may hear the remark, "It's easier to lift the bucket with my arm than with the pulley". This may be the case. However, the pulley may allow children to lift something to a height which would otherwise be impossible for them to achieve.

Be very careful to ensure other students are well away from the lifting area, as a bucket of water dropped from a metre or two would be dangerous.

pulleys 2 Depending on the type of pulleys you are using, you may need to use some extra rope or wire to keep the moveable pulley in the correct position. Be sure to practice this one first.

pulleys 3 It is very important to keep your technique accurate in this experiment. You should get very good results which show that for the double pulley, the rope is pulled

twice as far. This is where the ease of lifting comes from. The rope is pulled with half the effort, but twice the distance.

 inclined planes

ramps In carrying out scientific experiments, learning to follow the method is a critical first step for students. The aim of this lesson – in addition to demonstrating the use of a ramp – is to learn the importance of repeated trials of a procedure.

The distance the car rolls from each height should be repeated three times, and the distances then averaged. If some children are not able to average and graph results, still have them do the repetitions then in your review, ask their opinion about the outcomes, (e.g. "This one only went a short distance because it fell off the edge, so maybe it wasn't a good measure of how far really the car would travel", or "All the distances were nearly the same", etc.). The ramp serves to make it easy to move from one height to another.

You should have found that the car rolling from a plank with a steeper slope (greater height) rolled furthest. They should attribute this to the greater potential energy of the car at the start (refer to gravity 1).

wedges When effort is applied to the thick end of a wedge, the force is transferred to the sloping sides. Then the force is transferred to the wood, stone, etc., causing it to split.

When gluing on the ramps to make the wedge, one must be turned over to make it work.

The best zippers to look at are large open-ended zippers, like on a jacket. You will see one side of a double-sloped edge. This forces the teeth apart when it is pulled down. To close the zippers back up, there are ramps on either side of the zipper slide, which push the teeth back together. Nails usually have four sides on the point. Other wedges include:
– knife (one or two sides)
– chisel (one or two sides)
– needle (completely rounded – many sides)
– saw: each tooth is a wedge (one or two sides).

screws 1 This exercise shows that a screw is really a "rolled-up" ramp. As we trace up from the bottom we find our finger moving up the pencil along a spiral path. These "screws" can be used as decorations for your room if done with coloured paper and markers.

screws 2 Many machines which we see everyday use the principle of lifting and lowering something on a screw. Bottles of soft drink have caps which operate this way. They make use of the force of friction to tighten the cap down after it is opened. Small drills mine sawdust out from the surface of the wood. Large drills can mine earth upwards at a building site. Swivel chairs and stools adjust your height using a large screw. There is not much friction applied as the seat can usually be spun around freely by the occupant.

 ## levers

levers 1 You should notice that the counter flips highest when the effort is closer to the fulcrum, and the fulcrum further from the load.

In the extension the heavy load can be lifted easily if it is close to the fulcrum and the effort (or person lifting the others) is far from the fulcrum. Mention that different types of levers are for different purposes. In the first experiment, the objective was to gain maximum movement and the effort is placed closer to the fulcrum from the load. The objective in the extension is to gain maximum force so the effort is farther from the fulcrum than the load (see diagram on page 49 and 50).

levers 2 You can move some bricks or bags of sand if your children are not big enough or strong enough to move each other in the wheel-barrow safely. The load should be placed in the centre of the wheelbarrow or towards the front wheel. If the weight is too far back towards the handles, the wheelbarrow can tip backwards. When this happens, the fulcrum has been shifted towards the handles.

levers 3 Chopsticks can be the disposable wooden type. Noodles can be macaroni, or shells, or something easy to pick up. Don't use long spaghetti noodles, as they will be tricky to pick up. Your jaw is the other lever you use. As an extension, some students may be interested in finding those facial muscles and drawing and labelling an accurate diagram.

Other third class levers are your arm (when you throw and use a racquet), or your leg when you kick a ball; also your arm and hand combine to form a fulcrum for a shovel.

levers 4   tongs – class 3   garlic press – class 2
            scales – class 1   scissors – class 1
            stapler – class 3   nutcracker – class 2

levers 5 Have stations set up in separate areas.

levers 6 Be sure not to put the holes near the edge or they will rip. The ears' holes can also be made stronger by reinforcing the holes with tape or reinforcements.

The string should be long enough to be reached by your hand when held down at your side. This is two third class levers operated at the same time by the same effort.

 ## tools

tools These outcomes may vary from model to model.

tool	type	no.	contribution
eggbeater	gears & axles	3	The large gear turns the two smaller gears in opposite directions. Blades are attached at the end of the axles of the small gears.
ice cream scoop	gear & axle	2	The blade on the end of the gear axle is curved to scrape out the ice-cream from the scoop. The plunger returns to its original position by a spring attached to the screw.
	screw	1	screw holds spring.
pizza cutter	wheel & axle	1	wheel rolls blade along.
	wedge	1	The wheel's edge is a wedge to cut.
	lever	1	The fulcrum is at the axle and the effort is on the handle. The load is the wheel and the pizza it cuts.
corkscrew	screw	1	secures the cork so it can be pulled.
	levers (2nd class)	2	moves the gears (the load).
	gears	2	moves the central shaft upwards, bringing the cork out

 ## projects

It is a good exercise to do some research on the drawings of Leonardo's machines. Find out how the machines were to be used, and whether or not they would have actually worked.

Some of the machine drawings have been built, and an exhibition of Leonardo's machines constructed by Artisans of Florentine – Teknoart, based on Leonardo's famous codices, has been on display internationally.

For further help and information and details of our Primary Physics series, visit our website:
www.primaryphysics.com

answers

quiz

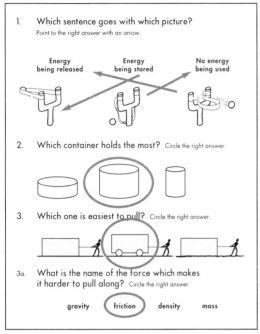

1. Which sentence goes with which picture?
   Point to the right answer with an arrow.

   Energy being released — Energy being stored — No energy being used

2. Which container holds the most? Circle the right answer.

3. Which one is easiest to pull? Circle the right answer.

3a. What is the name of the force which makes it harder to pull along? Circle the right answer.

   gravity · friction · density · mass

quiz (continued)

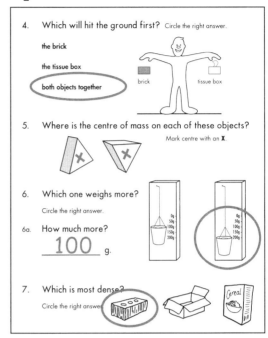

4. Which will hit the ground first? Circle the right answer.

   the brick

   the tissue box

   both objects together

   brick   tissue box

5. Where is the centre of mass on each of these objects?
   Mark centre with an **X**.

6. Which one weighs more?
   Circle the right answer.

6a. How much more?
   100 g.

   0g
   50g
   100g
   150g
   200g

7. Which is most dense?
   Circle the right answer.

   Cereal

puzzle

balance   friction
centre of mass gravity
density   mass
energy

e nergy

g ravity

f orce

m ass

f riction

b alance

d ensity

c entre of mass

Notes

Question 3
The answer is **friction**. The wheels reduce the frictional force, and allow the block to move easily. In a discussion you may wish to add that the ability to roll is actually dependent on friction.

Question 4
The objects have been drawn so that air resistance is not a factor.

Question 5
The position of an object does not affect its centre of mass.

equipment list

Start saving these things up in advance:

STATIONERY
- butcher's paper
- cardboard, various
- chalk
- craft glue
- crayons
- elastic string
- glue stick
- graph paper
- hole puncher, single
- marking pen
- matchsticks (headless)
- paper
- paper, waxed
- paper, recycled
- paper clip
- paper fasteners (brads)
- pencils, coloured
- pencils, various
- poster boards (in shapes)
- rubber bands, assorted
- ruler, 30 cm (12")
- scissors
- stapler
- staples
- stickers (foil stars)
- sticky tape
- string

CONTAINERS
- containers (e.g. 500 ml, 1 litre container of juice)
- containers, plastic e.g. 250 ml, 500 ml, 1 litre (half pint, pint, quart)
- containers, small, x 2 (e.g. 250 ml (8 oz) fruit cups)
- cup, plastic
- large cardboard box
- lid, plastic 12 cm (5") dia.
- lids, metal, equal sized x 4
- lots of empty drink cans, cartons, milk cartons or blocks of same size
- measuring containers, different sizes
- metal lids, 7 cm (2 3/4") dia.
- plastic bag (e.g. shopping bag)

WORKSPACES
- climbing bars (or monkey bars)
- flat space to use a wheel-barrow
- floor, smooth
- sandpit
- table

HARDWARE
- axe
- bolt & nut, 50 mm x 2 mm (2" x 1/16") dia.
- bolt, 5 mm (1/8") dia.
- brick
- chisel
- electrical tape
- hammer
- measuring tape
- nails, various
- nut and bolt, large
- planks
- pulleys x 3
- rope, 6 m (6 yds)
- sandpaper, different grades 6 cm x 8 cm (3" x 5") ea.
- scissors
- shovel
- tape measure
- toilet rolls, empty
- washers
- washers, 2.5 cm (1") dia. x 5
- weights
- wheel-barrow
- wood, various sizes
- wooden block, 8 cm x 10 cm x 20 cm (2" x 4" x 8")
- wire, 1 mm dia.
- wire cutters

FOOD
- apple
- banana
- beans, dried
- bread, white
- bread, pumpernickel
- cooked rice
- dough
- egg (hard boiled)
- flour, white
- hard-boiled egg
- marshmallow
- noodles (e.g. macaroni, shell, etc.)
- onion
- orange
- potato
- rice
- sugar
- tinned food, various weights (50g, 100g, 200g)
- walnuts (in shell)
- water

HOUSEHOLD ITEMS
- aluminium foil
- balance scales
- bathroom scale
- bicycle (with gears)
- book, hardcover
- bottling tops ("crown" bottle tops)
- brick
- broom/broomstick
- buckets (plastic & metal)
- button (flat)
- chair
- coat-hanger (wire)
- coins
- chopsticks
- clothes pegs
- corkscrew
- cotton spools, empty
- cotton wool
- dolly peg
- egg beater
- garlic press
- ice cream scoop
- large zipper (open-ended)
- newspaper
- nut cracker
- paddlepop (popsicle) sticks
- pizza cutter
- plates
- rope, short pieces
- scales
- sponge, natural
- sponge, thin
- swivel chair (not pneumatic)
- table, folding
- tissues
- tongs

TOYS
- blocks
- counters
- cuddly toy
- golf ball
- marbles (regular & large)
- piñata
- plasticine (modelling clay)
- sand
- sea shells
- see saw
- tennis raquet & balls
- toy car
- gliders, cardboard or balsa

conclusion

Leonardo da Vinci looked closely at the world around him and made careful observations from which he could see the principles behind how things worked. He used this understanding to design machines unknown in his day. He made his own investigations and so pioneered the scientific method.

By doing the exercises in this book, you have begun to experiment and look for yourself at the way things work. Let it be the beginning of a new way of looking at the world. At every opportunity, observe, question, and be curious. You might even make drawings and models to show your thoughts, just as Leonardo did.

It is our hope that the time spent doing these hands-on lessons has provided you with the confidence, encouragement and inspiration for real learning. Take the skills you have practiced here out to your environment and use them. The end result will be that the joy and excitement of the spirit of learning becomes your own.

For other *Primary Physics* books, visit us on the web at: **www.primaryphysics.com**

About the author

Marti Ellen has a Master of Science from Stanford University, USA, a Dip. Ed. from the Sydney Institute of Education and is the author of several papers on marine biology. She has taught physics to young children since 1985.

About the illustrator

Andrew Davies is a graphic designer and illustrator. He has a Bachelor of Design in Visual Communications from the University of Technology, Sydney and is the author and illustrator of *The Big Brain Book* (HarperCollins, 1993).